Paulo Tebulo
Adelina Adelino
Banu Irénio

Botanical Insecticides for the Control of(Alocypha bimaculata Jacoby)

Paulo Tebulo
Adelina Adelino
Banu Irénio

Botanical Insecticides for the Control of(Alocypha bimaculata Jacoby)

on the Yield ofSesame (Sesamum indicum L.) in the Edaphoclimatic Conditions of the Mapupulo-Cabo Delgado Agronomic Post

Imprint
Any brand names and product names mentioned in this book are subject to trademark, brand or patent protection and are trademarks or registered trademarks of their respective holders. The use of brand names, product names, common names, trade names, product descriptions etc. even without a particular marking in this work is in no way to be construed to mean that such names may be regarded as unrestricted in respect of trademark and brand protection legislation and could thus be used by anyone.

Cover image: www.ingimage.com

This book is a translation from the original published under ISBN 978-620-6-76135-8.

Publisher:
Sciencia Scripts
is a trademark of
Dodo Books Indian Ocean Ltd. and OmniScriptum S.R.L publishing group

120 High Road, East Finchley, London, N2 9ED, United Kingdom
Str. Armeneasca 28/1, office 1, Chisinau MD-2012, Republic of Moldova, Europe
Printed at: see last page
ISBN: 978-620-8-20488-4

TABLE OF CONTENTS

Summary

The aim of this work is to evaluate the performance of botanical insecticides in controlling the leaf beetle (*Alocypha bimaculata* Jacoby) on the yield of the *sesame* crop (*Sesamum indicum L.*) in the soil and climate conditions of the Mapupulo agronomic post, Cabo Delgado Province. An experiment was conducted in the 2022/2023 campaign in the fields of the Mapupulo Agricultural Research Center (CIAM), which belongs to the Mozambique Agricultural Research Institute (IIAM). A Causalized Complete Block Design (DBCC) was used, consisting of 4 blocks and 5 treatments based on Margosa, Castor, Eucalyptus, Control and Tobacco. The following entomological parameters were observed: herbivory index before and after application, number of dead pests, number of live pests, and agronomic parameters: number of pods per plant and pod weight and yield in kg/ha. In terms of the efficiency of the botanical insecticides, the Tobacco extracts were observed and highlighted as having the lowest rate of pest attack as herbivory in their plots, followed by Margosa extracts, so that the control had the highest number of pests and damage. Excel was used to construct graphs and tables. The statistical package used was Statistix 10.0 for ANOVA and based on Tukey's test at 5% significance level, significant differences were obtained in all variables, so that in the yield variable the Tobacco extracts obtained an average of 1501.0 kg/ha and the Margosa extracts had an average of less than 992.5 kg/ha.

Keyword: *Sesamum indicum L. Alocypha bimaculata* Jacoby, botanical insecticides.

Abstract

This study aims to evaluate the performance of botanic insecticides in controlling the leaf beetle (*Alocypha bimaculata* Jacoby) in the sesame crop (*Sesamum indicum L.*) under the edaphoclimatic conditions of the agronomic station of Mapupulo, Cabo Delgado Province. An experiment was conducted during the 2022/2023 campaign in the fields of the Agricultural Research Center of Mapupulo (CIAM), which belongs to the Agricultural Research Institute of Mozambique (IIAM). A Randomized Complete Block Design (RCBD) with 4 blocks and 5 treatments was used Tabaco, eucalyptol, Neen, and ricin. The following entomological parameters were observed: herbivory index before and after application, number of dead pests, number of live pests, and agronomic parameters: number of pods per plant, pod weight, and yield in kg/ha. In terms of biopesticide efficiency, tobacco, extracts had the lowest pest attack index, such as herbivory in their plots, followed by neem extracts, while the control had a higher number of pests and damage. Excel was used for the construction of graphs and tables, and the statistical package used was Statistix 10.0 for ANOVA. Based on the Tukey test at a 5% significance level, significant differences were observed in all variables. In terms of yield, tobacco extracts had an average of 1501.0 kg/ha, while neem extracts had an average of less than 992.5 kg/ha.

Keywords: *Sesamum indicum* L, *Alocypha bimaculata* Jacoby, botanic insecticides

List of abbreviations

CIAMMozambique Research   Center

            CVCefficient of Variation
m²   Square meters
            GGrams
FAO   Food and Agriculture organization
IIAM   Agricultural Research Institute of Mozambique

            Hectares
MAE   Ministry of StAdministration
            KmKilometers
            MmMillimeter
MADER   Ministry of Agriculture and Rural Development

            SDAEServiços Distritais de Actividades Económicas
            ANOVAnalysis of Variance
MASA   Ministry of Agriculture and Food Security

 ABJ   *Alocypha bimaculata* Jacoby

USAID United State Agency for International Development

CHAPTER I. INTRODUCTION

1.1. Sesame cultivation in general

Sesame (*Sesamum indicum* L) is known as one of the oldest oilseeds produced worldwide and originated in Africa, where most of the wild species of the *sesamum* genus are concentrated (Sousa et al., 2017).

Agriculture plays a vital role in food security, employment and socio-economic development in many countries, but it is constantly threatened by pests and diseases that can cause significant damage to crops, leading to substantial losses in production. The leaf beetle (*Alocypha bimaculata* Jacoby) is a pest that affects the sesame crop, causing severe damage to the leaves and harming crop yields (IIAM,2020).

In this context, the use of botanical insecticides has been highlighted as a sustainable alternative for controlling agricultural pests, reducing the damage caused by insects and minimizing environmental impacts and exposure tharmful chemicals. This study aims to evaluate the performance of botanical insecticides in controlling the leaf beetle and its impact on sesame crop yields in Mozambique (Guimarães, 2012).

The problem lies in the decrease in the quality and productivity of agricultural products due to leaf beetle infestation, especially in areas of small producers who often do not have access to inorganic pesticides due to financial constraints. The intensive use of chemical products has adverse effects on the environment and human health, making it necessary to look for safer and more sustainable alternatives, such as biopesticides (Patrícia Garcia Ferreira, 2022).

According to the Instituto de Investigação Agraria de Moçambique - IIAM (2020), the average yield of the Lindi sesame variety is around 800 kg to 1000 ton/ha, varying from region to region.

In the Montepuez district, sesame cultivation is practiced in the family sector by small producers, under rainfed conditions. The district of Montepuez is centered in the R7 agro-ecological region, with an average yield of 800kg/ha.

1.2. Problem definition

In recent years, the appearance of the leaf beetle pest (*Alocypha bimaculata* Jacoby) in the country, particularly in the district of Montepuez, has been responsible for the decline in the quality and productivity of products. Since most dthe Mozambican population practices family farming, their incomes have been low, as most producers lack the financial resources to obtain inorganic pesticides.

Small producers choose to use chemical products, sometimes in high doses, which in addition to being harmful to the environment can lead badverse consequences such as the appearance of resistant organisms and intoxication of the plant itself.

However, the use of chemical products is something that has been expanding even today and with more intensified effects, these products are cars and when used intensively and incorrectly they create damage to the environment as well as to man. With this in mind, the aim of this study is to answer the alternative use of bioinsecticides that are better, slightly toxic and contribute to nutritional quality, for the control of the leaf beetle (*Alocypha bimaculata* Jacoby*)* without harming the environment and the consumer extracted from plants (tobacco, eucalyptus, castor and margosa) and help increase yields?

1.3. Justification

As usual in Mozambique, sesame is seen as a flagship source of income for Mozambican families, and it also contributes significantly to the commercial value of the region and the country as a whole. In the same vein, producers are called upon to put in a certain amount of effort and effort to increase production and productivity in order to meet their needs and those of the country as a whole.

In the district of Montepuez, most of the producers grow sesame, which is a source of food as well as income. It is an indispensable crop due to its nutritional content and its diverse potential, and the increased use of bioinsecticides will increase productivity and reduce environmental pollution.

This study deals with a practical, easily accessible and low-cost approach that comprehensively addresses environmental aspects by creating a sustainable, conversational space, i.e. the application of botanical insecticides.

As a social gain, the small producer will be equipped with a new production technique, in the form of control
by means of bioinsecticides that result in an increase in family income and wellbeing academically, it is seen as a sustainable framework that covers yesterday, today and tomorrow;

However, this study is expected to show signs of improvement in the low yields caused by beetle attacks, which growers have been complaining about in this region.

1.4.   Objectives:

1.4.1   General objective

To evaluate the performance of bioinsecticides in controlling the leaf beetle (*Alocypha bimaculata* Jacoby) on the yield of the *sesame* crop (*Sesamum indicum L.*) in the soil and climate conditions of the Mapupulo agronomic station.

1.4.2. Specific objectives

✓ Check the level of pest incidence in the different treatments before and after application of the botanical insecticides.

✓ Comparing the efficiency of botanical insecticides in controlling the leaf beetle

✓ Describe the growth parameters (damaged leaves and number of live and dead pests) and yield (weight of capsules, number of capsules) when botanical insecticides are applied;

✓ Relate the pest incidence rate after application to the characteristics of the growth and yield parameters.

1.5. Hypotheses

H0: Botanical insecticides are not efficient in controlling the leaf beetle (*Alocypha bimaculata* Jacoby).

H1: At least one of the botanical insecticides was effective in controlling the leaf beetle (*Alocypha bimaculata* Jacoby).

CHAPTER II. BIBLIOGRAPHIC REVIEW

2.   Description of the sesame crop

The sesame crop originated in Africa and, despite being a crop of great economic value, its cultivation is still very limited to small areas. It is a very old crop and is considered one of the main oilseeds produced worldwide, ranking ninth among the most produced cereals (Queiroga et al., 2008).

2.1   Taxonomy of the sesame crop

The sesame crop is a plant that belongs to the kingdom plantae, division, magnoliophyta, class magnoliopsida, Order lamiales, Family Pedaliaceae, genus *Sesamum* and Species: *Sesamum indicum* (Arriel, 2009). The crop is considered to be a predominantly autogamous species, so some authors report different rates of allogamy that vary from region to region (Firmino et al., 2001).

The sesame crop has an erect stem with a pivoting root system, its main characteristic is its heterogeneity in terms of the number of flowers and fruits per leaf axil, fruit size, seeds, its descending and indehiscent characteristic, the seeds are small usually weighing 2 to 4 g depending on the variety and the environment, the color of the seeds varies from white to black which can have a direct influence on their grain yield and oil content (Beltrão et al., 2007).

The leaves are alternate, and those on the lower part of the adult plant are broader, irregularly toothed or lobed, while those on the upper part are lanceolate. The flowers are complete and axillary, varying from 1 to 3 per leaf axil. The fruit is in the form of an elongated capsule, hairy, dehiscent when it reaches maturity and opens and scatters the seeds or indehiscent (Embrapa, 2007 cit by Arriel et al).

2.2. Importance of culture

Sesame is a crop of great economic value and can be a source of income for most producers, particularly in the family sector. Its consumption is indispensable due to the benefits it brings to human health in preventing different types of diseases (Arriel, Beltrão & Firmino, 2009).

Sesame seeds are rich in protein and are used to make bread, cookies and sweets. They also have a high oil content, around 50%, which can be used in the food and chemical industries, as a herbal medicine, cosmetics and animal feed (Correia et al., 1995; Arriel et al., 2006; Queiroga et al., 2008). Around 70% of sesame grain production is destined for processing to obtain oil and food products (PERIN et al., 2010).

Currently, several authors report that the consumption of sesame is important because it brings benefits to human health, helping to prevent various diseases such as depression, osteoporosis (because it is rich in calcium), cholesterol (lecithin) and atherosclerosis (Queiroga, Arriel & Silva, 2010).

2.2.1. Cooking

Sesame oil for human consumption has a long history and varies according to the region and customary practices of the people who extract it. The use of sesame seeds in industry involves the manufacture of bran for feed, flour, pies and confectionery (Godoy et al, 1985).

2.2.2 Cosmetics production

Sesame oil can easily penetrate the skin, nourishing and detoxifying the deeper layers of tissue (Morrison & Foerster, 2008).

2.2.1 Pharmaceutical action

Sesame leaf tea is used to relieve diarrhea, the oil from the seeds is used as a poultice for burns, it is considered to be a stimulant for the release of breast milk, as well as being known to have an anti-rheumatic and anti-inflammatory action, in Asian countries it is used to treat wounds, mainly caused by burns (Alvarez et al., 2008).

2.3. Cultivation

In Mozambique, sesame is grown in the northern and central regions of the country, mainly by small-scale producers, and cultivation is important so that the crop can express its maximum production potential, productivity and profitability (Embrapa, 2009).

2.3.1. Soil preparation

The sesame crop is a very slow-growing plant, which is critical in the first 45 days after plant emergence. During this time, it must be kept free of weeds, so soil preparation is a very important tool, because it ends up working as a weed control method, and it is recommended that it be done two or three times with a hoe or tiller before sowing (Embrapa, 2007).

When preparing the soil, whether mechanized or by hand, it is important to use agricultural implements appropriately due to tedepth, relief, degree of structure and structural class of the soil, always avoiding clods of soil, due to the nature of the seed as it is small and light, preferring an even bed for good germination (Queiroga, Arriel, and Silva, 2010).

2.2.1 Sowing and germination

Sowing can be manual or mechanized, depending on the area farmed and the technological level of the crop, with the amount of seeds per hectare varying between 1.5 kg and 3.5 kg, depending on the spacing and density of the planting. Sowing can be done in shallow furrows or in shallow pits with a depth of 2 cm (Araújo, 2005).

A spacing of between 0.60 cm and 0.80 cm between rows and 0.10* 0.20 cm between plants is recommended for unbranched varieties. For planting in double rows, a spacing of 1.70 cm between rows and 0.10 cm between plants is recommended (Beltrão et al., 2001).

Sesame is a slow-growing plant at first, with critical growth in the first few days after seedling emergence, which in turn takes between 3 and 5 days depending on the variety and different soil and climate conditions, and must be kept free of weeds and pests (Embrapa, 2009). However, when sesame plants are subjected to nutritional stress, the capsules containing the seeds may abort (Beltrão et al., 2001).

2.3 Soils for growing sesame

The soils best suited to sesame production are loamy, sandy loam, well-drained soils with a minimum depth of 60 cm, with natural fertility, the crop does not tolerate Ph below 5.5 or above 8, it is very sensitive to salinity and alkalinity (IIAM, 2020).

Júnior & Azevedo (2013) state that the sesame crop can grow and develop in any type of soil as long as they are well-drained and have a good structure. Clay soils and soils with compaction problems should be avoided because the oxygen in the roots can be limited to the point of compromising the growth and development of the plants and their economic production capacity.

2.2. Weather

The crop is widely adaptable to the soil and climatic conditions of hot tropical climates and is tolerant to water stress (Beltrão *et al.*, 2010). However, the main climatic factors that determine a good development of the crop in terms of temperature range from 25 to 30 C, with a minimum rainfall of 300 mm, the highest luminosity of the crops is 10 hours of light per day, and altitude preferably low altitudes and up to

1,200 m (Embrapa, 2009).

2.3. Temperature

The sesame crop does not thrive at very high altitudes and as a result the plants become stunted, without much branching and with low production (Embrapa, 2014). Temperatures below 20 ^{o}C, 10 oC and above 40 ^{o}C are not favorable, as there is a delay in the process of germination and plant development, paralysis of cell metabolism and abortion of flowers with or without grain filling (Silva, et al., 2019).

However, average temperatures of $27°C$ are favorable for the vegetative growth and ripening of sesame pods.

The increased yield of the sesame crop depends on rainfall, which is distributed as follows: at germination the crop needs 35%, at flowering 45% and 20% at the beginning of fruit ripening (capsule). In places with rainfall of less than 300 mm, the crop can produce 300 to 500 kg/ha of grain. The water requirement of the sesame crop is directly related to the distribution of the total amount of rainfall during the vegetative phase of the plant. Although this crop is resistant to drought, it is sensitive to waterlogging of the soil, i.e. soil moisture is beneficial for flowering and fruiting, but heavy rainfall causes flower abortions and lodging of the plants (Firmino et al., 2007). The optimum range for sesame production is between 500 and 600 mm (Grilo Júnior & Azevedo, 2013).

2.2 Nutritional requirements of the sesame crop

For the sesame crop, the most limiting factor **is** nitrogen in terms of nutrients at the time of production, as it is responsible for important functions in plant metabolism and nutrition. Its deficiency causes a nutritional disorder with a drop in production and its increase causes an increase in the incidence of pests and diseases, a drop in production and a reduction in oil content (Biscaro *et al.,* 2008).

Souza *et al.* (2014) verified the initial development of sesame with nitrogen fertilization, and concluded that fertilization with 85.6 to 119.2 kg ha of N promotes good plant development, such as height and stem diameter.

One of the ways to reduce losses and production costs is to useorganic inputs, which are often easily found on most rural properties (Silva et al., 2010). From the same perspective, Monteiro, (2014) reports that the use of organic products such as extracts of margosa, cow milk, tobacco, molasses, goat manure, castor, can be used in the constitution of organic nutrient solution and increase in crop yield, and also in reducing the use of inorganic inputs, reducing production costs.

Phosphorus (P), on the other hand, is also an essential nutrient required by the crop. In the case of sesame, this nutrient is involved in metabolic processes such as nutrient absorption and the formation of the plant's different organs (Frandoloso, 2006). Prado (2008) says that potassium (K) is considered a fundamental element for growth, development and grain quality.

Carneiro *et al.* (2014) evaluated organic and phosphate fertilization in sesame cultivation, and found that organic fertilization can replace chemical fertilization to meet the crop's needs, having a similar or even superior effect to chemical fertilization.

2.2 Harvesting

Depending on the environmental conditions and the variety, sesame has a cycle that varies between 3 and 4 months. When it comes to harvesting, care must be taken as the fruit opens naturally when ripe. The crop should be harvested at the moment of physiological ripeness, i.e. from the moment the plants begin to turn yellow on the branches and flowers, and the basal capsules begin to open. Harvesting can be done by hand or mechanized depending on the area being farmed (Antoniassi et al., 2013). However, fruit ripening does not occur uniformly, as the capsules at the base of the plant open earlier, demonstrating the exact time to start harvesting. If the crop is harvested later, there will be a greater loss of grains, as fruit dehiscence progresses rapidly (Lagos, 2001).

For manual harvesting, the most commonly used technique is to cut off the base of the plants at the point where the first fruit is inserted. After cutting, the bundles are tied together, leaving the apexes facing upwards. They should be joined together and laid out for natural drying, preferably in the same place. After drying, they are beaten on a plastic sheet to prevent the seeds from scattering. If it is necessary to continue drying the grains after beating, it is advisable to spread a thin layer of grains on the plastic sheet until the humidity is between 4 and 6%. Care should be taken especially with the wind, which can contribute to dry capsules falling to the ground (Silva, 2010).

2.2. Income

The average yield of sesame is around 650 kg/ha, but up to 1,500 kg/ha depending on the variety and with adequate fertilization (Neto *et al.*, 2016). It should be noted that this can vary depending on the soil, climate, variety, plant populations, cultivation, number of seeds per capsule and seed weight, with some genotypes yielding up to 1,800 kg/ha (Embrapa, 2009).

2.3. International production

According to FAO (2019), the country that produced the most sesame was Sudan, with a volume of 1,210,000 tons/ha, contributing 18.47% to global production. Sesame is produced in 71 countries on the Asian continent and in Africa. World production is estimated at 3.16 million tons, and the cultivated area is 6.56 million hectares with a yield of 481.40 kg/ha. India and Myanmar are responsible for 49% of world sesame production, and Brazil is also one of the smallest producers of this crop, with 15,000 tons produced on 25,000 hectares, and with a yield of around 600.kg/ha (FAO).

2.3.1. Production in Mozambique

From an economic point of view in Mozambique, sesame has stood out on the national scene, showing a slight upward trend in terms of production, but still very low compared to other African countries. Mozambique is the world's ninth largest producer with a production of 69,352 tons, and in the 2019/2020 agricultural season, the country recorded a figure of 118,402 tons of sesame (MADER). Demand for the product is very high, and it has reached rewarding prices on the domestic and international markets (Cruz et al., 2010).

Table 1: Income from sesame in the family sector in the Montepuez district Campaigns Yield obtained (ton/ha)

2018/2019	0.2
2019/2020	0.3
2020/2021	0.2
2021/2022	0.4
2022/2023	0.2

Source: SDAE-Montepuez

The yield of the Lindi sesame variety can reach 800 kg per hectare in rainfed production, but due to pest attacks these yields may not be achieved (Malta, Neto, Silva, Rocha, Castro, Reis and Akerman, 2016).

2.2. Pests

They are organisms that reduce the production of crops grown in the field by infesting them and being vectors of diseases, drastically reducing the productivity and yield of the final product when not controlled before causing major damage to agricultural production (Picanço, 2010).

2.2.1. Main pests of the sesame crop

The main pests of the sesame crop are: The reel caterpillar (*Antigastra cataunalis*), aphid (*Aphissp*), leaf beetle (*Alocypha bimaculata* Jacoby), green leafhopper (*Empoascasp*) (saúvas (*Attaspp*), whitefly (*Bemisia argentifolii*).

2.2.2. Brief history and description of the pest under study

The genus Jacobian beetle was first described by Maulik in 1926 with the species Sphaerophy sapiceicollis Jacoby in 1889n from Burma. Later Chen in 1934 attributed to these two genera two taxons, Jacobianapiceicollis and jacobyana nigrofasciota, later scher 1969 described the same species, and recently Jacobyana was found in Panchther, Nepal in 2001, finally in India (Biondie & D'Alessandro, 2011).

The taxonomic classification of *Alocypha bimaculata* Jacoby according to Sidumo, (2006, cit. in Mucavea, 2010): kingdom: animalia, class: insecta, order: coleoptera, family: chrysomelidae, subfamily: alticinae or heliticinae, genus: *Alocypha*, species: *Alocypha bimaculata Jacoby*.

2.3.2 Leaf beetle

(*Alocypha bimaculata* Jacoby) according to its classification, this species belongs to the coleoptera family, chrysomelidae being one of the largest families including approximately

37,000 to 40,000 species (Jolivet & Verma, 2002). It is considered to be a group of herbivores, largely related to its host plant and its abiotic characteristics (Linzmeier & Ribeiro-costa, 2012).

The **leaf beetle** (*Alocypha bimaculata* Jacoby), commonly known as the flea beetle, originated in Africa and can currently be found widely in Malawi, southern Tanzania and Mozambique. It is considered one of the pests that most limits production, especially in the first stage of the crop, generating losses in production and productivity in both the large and adult stages (Palote, 2017).

2.11.3. Description

According to Biondi & D'Alessandro (2006) the leaf beetle (*Alocypha bimaculata* Jacoby) is described as follows:

2.11.4. Holotype

The sesame beetle has a black dorsal integument with a shiny or metallic appearance, a reddish apex of the elytra, with a rounded body (LB = 2.48 mm), strongly convex. Maximum pronotal width at the base (WP = 1.28 mm); maximum width of the elytra in the basal quarter (WE = 1.71 mm).

2.11.5. Border and vertex

It has a distinctly greenish, finely dotted surface, with different septal punctures, with subtriangular, brownish, poorly delimited frontal tubercles, with a fuzzy surface; distally deep frontal furrows, particularly along the ocular margin; interantennal space distinctly narrower than the length of the first antennomere, medially with two not well delimited setiferous pores; frontal carina not elevated; clypeus triangular with large setiferous perforations, lip sub-rectangular, distally brownish, palpus yellowish, eye subelliptical, of normal size, has antenna much shorter than body length, totally pale, but with antennomeres ranging from 5 - 11 slightly overshadowed, (right antenna).

2.11.6. Leg

They are entirely reddish-brown in color with a partially blackened femur; a slightly curved hind tibia without a toothed outer margin, and a short, reddish apical spur of the hind tibia.

First anterior and middle tarsomeres very slightly enlarged, with adhesive bristles on the ventral side

2.11.7. Ventral surface

The ventral part is black, with dense and fairly evenly distributed setiferous punctures, medially more sparse or absent on the prosternum, metasternum and the last four visible abdominal sternites, the last abdominal sternite without special pre-apical impressions.

This beetle in its adult stage as well as its larvae are considered herbivores and can feed on stems, leaves, roots, and flowers in almost all families of higher plants such as: Solanaceae, Brassicaceae, Lamiaceae, Pedaliceae (Rech, 2018).

The eggs are pale yellow and the larvae are white with a dark brown, comma-shaped head on the last instar (Sidumo, 2006).

Olmi (2006) says that this species of beetle acts as a pest in the larval stage as well as in the adult stage, showing symptoms on the plant such as: perforated leaves in the first stage of the crop, i.e. after the emergence of the crop, it attacks and creates perforations in the leaf until it kills the plant.

2.4 Main diseases

The *sesame* crop, depending on the variety, soil and climatic conditions and the cultural environment provided, can be infested by various agents (bacteria, fungi, viruses) that cause diseases such as angular spot (*Cylindrosporium sesami*), which affects the leaf part, causing angular lesions. To control this disease, the use of resistant varieties is recommended (Embrapa, 2000).

2.4.1.   Phorid spot (*Cercospora sesami*) is usually caused by a fungus that attacks leaves and fruit, causing more or less regular rounded spots, the agent being transmitted by the seed. The fungus penetrates the inside of the capsule, reaches the seed and turns it black. The use of resistant varieties is recommended to control it (Arriel, et al., 2007).

Black stem rot (*Macrophomina phseolina*), a disease caused by a very dangerous fungus, mainly affects the stem and branches, causing light brown lesions, which can surround the same parts of the plant and even reach the plant's terminal bud, there may also be some black spots, but the affected plants wilt and even dry out and die, high temperatures, low soil humidity and low potassium availability favor the appearance of this pathogen. The disease is spread by water (rain or irrigation) contaminated by soil particles and infected seeds. It is recommended to treat with crop rotation and the use of healthy seeds, eliminating cultural remains and treating the seeds (Arriel, et al., 2007).

2.4.2. Fusarium wilt (*Fusarium oxysporium*) this disease is also caused by a fungus. The symptom is flaccidity and wilting of the plant, which then dries out and dies. It causes blackening of the tissues of the plant's vascular systems. The pathogen that transmits this disease survives in the form of spores in the soil, where it lives saprophilically on crop remains. It is spread by contaminated soil particles and water droplets (rain), and to control it we recommend using healthy seeds and resistant varieties (IIAM, 2020).

2.4.3. Bacterial wilt (*Xanthomonas campestri pv. semami*) At the beginning of the disease, rounded or angular dark spots appear on the leaves, stems and pods, which later turn reddish-brown or sometimes black and can join together to form a large necrotic area. This disease is spread by rainwater and wind, the transmission factor is the seed, excess nitrogen content in the field favors the appearance of this disease, for its control it is extremely important to eliminate plant remains and rotate the crop and finally use healthy seeds (Arriel, et al., 2007).

2.4.4. Altenaria blotch (*Altenaria sesami*) is an irregular circular brown spot on leaves and stems, which can bind and lead to necrosis of the affected area, causing defoliation and death of the plant. High temperatures favor the appearance of this seed-borne disease. As a control measure, it is recommended to rotate the crop, eliminate plant stubble and use resistant varieties (Arriel, et al., 2007).

2.4.5. Phylogeny

This anomaly is characterized by the shortening of the parts of the plant's stem located between two nodes (internodes), as well as the increased proliferation of flowers and branches in the apical region of the plant, which gives it a broom-like appearance. The modification of the floral organs into leaves renders the plant sterile. This anomaly is caused by Jassideos insects, the plants show a deficiency with leaf surfaces with areas of yellow color interspersed with areas of green color, to prevent its spread it is recommended to eradicate and burn the affected plants. (Arriel, et al., 2007)

2.2. Botanical Insecticides

There are a large number of plants in the world whose insecticidal activity has been studied. Plants from the Meliaceae, Rutaceae, Asteraceae, Annonaceae, Labiatae and Canellaceae families are considered the most promising (Jacobson, 1990).

In India, around 2,000 B.C., botanical insecticides (from plant extracts) were already used to control pests. In Egypt during the time of the Pharaohs and in China around 1,200 B.C. (Morreira, Picanço, Silva, Moreno, & Martins, 2007). In the last two decades, with the increase in pest resistance problems to organosynthetic pesticides, and the problems arising from the indiscriminate use of chemical insecticides on natural enemies, the environment, man himself and the development of organic farming, the search for organic food has increased significantly worldwide (Andrade & Nunes, 2001).

Along these lines, the term botanical insecticides (natural product), also called alternative, refers to products prepared from non-harmful substances, derived from plant extracts, which do not harm human health. These products can be used by humans, as well as the environment, to produce healthier food for the target group or end consumer (Iria, José, Gueds, Katell, & Sonia, 2020).

The use of botanical insecticides to control pests is a task for small African producers. Several products or alternatives have insecticidal properties that contradict some pests that mainly attack cereals in Africa, such as: Tobacco (*Nicotiana sp.*), *Tephrosia* (*Tephrosia vogelii*), margosa (*Azadirachta indica*), *Jatropha* (*Jatropha curcas*), garlic (*Allium sativum*), eucalyptus (*Eucalyptusspp*),

pyrethrins (*chrysanthemciner arifalium*), lemon (*citrus limon*), castor (*Rícinus communis*) (Stevenson *et al.* 2017).

2.2.1. Action of botanical pesticides on pests

Botanical derivatives can cause different effects on pests such as: repellency, inhibition of oviposition and feeding, changes in the hormonal system causing hormonal disturbances in development, deformities, infertility and mortality depending on the stage. The extent of the effects and the time of action depend on the dosage used, so bedeath occurs at high dosages with less intense and longer-lasting effects at lower dosages. The use of sub-lethals causes a long-term reduction in populations and requires smaller quantities of product. Lethal doses often make their use unfeasible due to the sheer quantity required (Roel, 2001).

According to Roel (2001), the efficiency of using botanical pesticides increases when the product is applied to smaller populations, with individuals at the beginning of their development. Depending on the plant species and the form of use, botanical pesticides can be used in pure form, as a powder, or as an extract (in aqueous solutions), as well as in other specific forms, conditions which facilitate monitoring and use.

For Cloyd (2004) the main advantages and disadvantages of using botanical insecticides are:

They degrade quickly, especially in conditions of high light, humidity, air and rain. They have low persistence in the soil as well as in the environment, reducing their impact on beneficial organisms, natural enemies, humans and the environment itself.

They are fast-acting and cause the insect to die or even reduce its feeding immediately after application.

They contain low toxicity to mammals, and some at the recommended dose are not toxic to humans, the environment as well as mammals.

They are selective, less harmful to beneficial insects mainly due to their low residual effect.

They are inexpensive to buy and most of them can be bought in our backyards until they multiply.

The use of plant extracts leads to less resistance in the development of pests, as heycontain active ingredients capable of inhibiting feeding and oviposition, growth, morphological changes, changes in the hormonal system and in the behavior of the insect (Galloet al.,2002).

2.2.1. Disadvantages of botanical insecticides

The same authors mentioned above point out the disadvantages of botanical insecticides:

It degrades quickly and requires many applications to achieve positive results in pest control;

• Many of the insecticides are not commercially available and are sometimes more expensive than inorganic insecticides (availability and cost).

• Natural insecticides have sometimes been less effective than synthetic ones;

• The results are not always immediate;

Bioactive compounds can differ in action according to the species and variety of plant, climatic elements such as light, temperature, relative humidity, rainfall and plant phenology (Shalaby et al., 1988; Russo et al., 1998; Andrade & Casali., 1999; Carvalho et al., 1999).

2.4.6. Description of the species used in the study

2.4.6.1. Castor or castor bean (*Ricinus communis* L.)

According to the taxonomic classification, the castor bean plant belongs to the Subdivision: Spermatophyta; Phylum: Angiospermae; Class: Dicotyledonous; Subclass: Archichlamydeae; Order: Geraniales; Family: Euphorbiaceae; Genus: *Ricinus*; Species: *Ricinus communis* (Milani, Junior, & Sousa, 2009).

Ricinus communis L. is considered to be a low-growing perennial oilseed, and a plant that has insecticidal activity often used as botanical insecticides, having originated in Africa (Milani, Junior, & Sousa, 2009). It is currently produced worldwide due to its easy adaptability, especially in tropical climates (Polito et al, 2019). The largest producers in tons in the world are: India, Beni, Iraq, Morocco, Togo, Mozambique, Cape Verde, China, Brazil, with India accounting for more than 90% of exports (FAO, 2020).

The leaves of the castor plant contain a low concentration of toxin (ricin), and can cause neuro-muscular problems when ingested. Generally, the symptoms of intoxication in animals begin to appear after a few hours or a few days, unlike the seeds which have a ricin metabolite, a very toxic substance harmful to animals and humans (Peron & Ferreira, 2012).

The plants of these species vary greatly in their characteristics, from leaf color, growth, oil content in the seeds, stems, and spiky and smooth fruits (Milani, Junior, & Sousa, 2009).

2.2.1.1. Margosa (*Azadirachta indica*)

Taxonomic classification of margosa or neem According to (Pio-correia, 1984)
Order: Rutales; Suborder: Rutinea; Family: Meliaceae; Subfamily: Melioideae;
Tribe: Melieae; Genus: *Azadirachta*; Species: *Azadirachta indica*.

Margosa or neem is a plant species that has insecticidal activity for controlling various pest species such as mealybugs, aphids, beetles and leafhoppers. It originated in India and spread to other continents (Viana, 2010).

The main compound contained in margosa is azadirachtin, authorized for foliar application on various crops such as corn, lettuce, coffee, citrus, papaya, melon, strawberries, peppers, tomatoes and cabbage, derived from different parts of the plant (Menezes, 2005). Azadirachtin is a substance that is very toxic to insects. Its repellent effect inhibits insect feeding and growth (Mordue & Blackell, 1993). The main sources of azadirachtin are fruit, leaves and bark (Neves, et al., 2008).

2.2.1.2. Tobacco (*Nicotina sp*)

Taxonomic classification according to (Reigart & Roberts, 1999). Phylum: Magnoliophyta; Class: Magnoliopsia; Order: Solanales; Family:

Solanaceae; Genus: *Nicotina;* Species: *Nicotina tabacum sp.*

Nicotine or tobacco is a plant with insecticidal action, by ingestion and mainly by contact. It was first used as an insecticide in 1690 in France in the form of a wash (Menezes, 2005). There are more than 15 species of the nicotine genus.

In the leaves we find alkaloids that grow in the lower region of the tip, and the nicotine content grows from the base to the tip of the main vein towards the edges, considered toxic to mammals, orally or even on the skin, where it is easily absorbed, nicotine is fast-acting, but is most active when applied in the hottest hours of the day, being completely degraded in 24 hours, leaving no toxic residue. It is not very toxic to plants (Marconi, 1988).

Nicotine is used as a botanical pesticide to control sucking arthropods such as aphids, whiteflies, leafhoppers, thrips and mites, mainly in greenhouse and cereal crops, with a contact and fumigation effect (Reigart & Roberts, 1999 cited by Cox, 2002).

In insects it acts on the nervous system with a very rapid effect, competing with acetylcholine by binding to acetylcholine receptors in the synapses of axons (Reigart & Roberts, 1999).

Nicotine spp is a very fast-acting toxin that acts on the nervous system of insects and also has contact and fumigation effects (Menezes, 2005). According to the same author, tobacco is considered to be the most toxic botanical insecticide used to control various pests such as aphids, bedbugs, cowbells, mealybugs and crickets on cereals and fruit plants.

2.2.1.1 Eucalyptus (*Eucaliptus sp*)

Eucalyptus belongs to the division: angiosperm, class: dicotyledonous, order: Myrtales, family: Mytaceae, genus: Eucaliptus, species: Eucaliptus sp. (Grattapagilia & Kirst, 2002). The eucalyptus plant is of Australian and Indonesian origin and has become one of the most widely produced forest genera worldwide due to its usefulness in the manufacture of wood and non-timber derivatives (Grattapagilia & Kirst, 2008).

There are more than 900 species recognized worldwide (Boland et al., 2006). They are trees that measure an average of 30 to 50 metres, varying from species to species (Brooker & Kleining, 2004).

Guerra (1985) states that eucalyptus can be used to protect grains against stored grain pests in general, and has insecticidal activity against *Triboliumcastaneum* (Herbst) beetles, Coleoptera: Tenebrionidae. The active substance contained in eucalyptus is called Citronellal or Eucalyptol.

CHAPTER III: METHODOLOGY

This study is explanatory in nature, as it aims to establish a cause and effect relationship by manipulating the variables relating to the object of study in a straightforward manner, seeking to find the causes of the phenomenon (Lakatos & Marconi, 2001). It is quantitative in nature, as it is characterized by the use of quantification, both in the way information is collected and in the way it is processed using statistical techniques (Richardson, 1998).

As for the procedure, it is experimental, i.e. it consists of determining an object of study, selecting the variables capable of influencing them and defining the rules for controlling and observing the effects that the variable produces on the object (Gill, 1999). With this, the study was carried out as a study factor which were the different biopesticides, based on plants such as: tobacco, castor, eucalyptus, and margosa in which we want to understand the efficiency that they may have had on the pests determined by leaf beetle (*Alocypha bimaculata Jacoby*) and to achieve this result an experiment was used with the aim of evaluating the cause and effect relationship making the harvest and the treatment to be ascertained according to the desired result be made based on measurements and statistical evaluations, seeking to answer the hypotheses mentioned above.

The study used botanical insecticides in their aqueous form to control the leaf beetle (*Alocypha bimaculata* Jacoby*)* in the sesame crop, and in order to achieve the aforementioned variables, it was decided to estimate the herbivory index, since the millimetre paper was not available and the estimation process was used for all treatments through the application of botanical insecticides.

Since this study included the use of botanical insecticides, in the case of the use of castor oil, a syrup was prepared using the leaves of the plant as an extract, they were crushed to prepare the syrup, after crushing, 250g was used for 1 liter of water, the remaining product was added with water in a container for 24 hours in a totally closed environment so that the active substance contained in the leaves mixed with water, then filtered.

In the case of nicotine, the first step was to defoliate 3 dry leaves, then place them in a container and soak them in 1 liter of hot water for 24 hours to mix the active substances with the water, then filter them and add them to 5 liters of water to spray an area of 15m² with attacked plants.

For eucalyptus, fresh leaves were used, which in turn were crushed with a pestle and mortar. For the formulation, 250g were used for 1 liter of water and then they were placed in a plastic container (bucket) and stored for 24 hours, and then went through the filtration process using a clean mosquito net, in the end obtaining a solution ready to be used in the field.

However, the margosa went through the same process of crushing the leaves, using a formulation of 250g of extract to 1 liter of water, leaving in an airtight environment so that the mixture occurs according to the active substances contained in the leaves for 24 hours, filtered after filtration, the soap was scraped and mixed and stirred until the mixture dissolved and had a homogeneous consistency.

3.   Materials and methods

3.1.   Delimiting the topic

➢   **Spatial:** the research was carried out in the province of Cabo Delgado, Montepuez district, at the Mapupulo Agricultural Research Center (CIAM).

➢   **Timing:** The research took place during the 2023-2024 campaign.

The trial was conducted at the Mapupulo Agricultural Research Center (CIAM) in the Montepuez district, located in the southern part of Cabo-Delgado province, belonging to the Mozambique Agricultural Research Institute (IIAM)

Northeast Zonal Center (Cznd), during the 2020/2023 agricultural season from February to June.

3.2. Geographical location of the study site

The district of Montepuez is located in the south of Cabo Delgado Province, 210 km from the city of Pemba, bordered to the south by the districts of Chiúre and Namuno, to the north by the district of Mueda, to the west by the districts of Balama and Mecula (Niassa Province), and to the east by the districts of Meluco and Ancuabe (MAE, 2005).

3.3. Edaphoclimatic conditions in the Montepuez district

3.3.1 Climate and relief

According to the Ministry of State Administration (2005), the Montepuez district is a region dominated by semi-arid and dry sub-humid climates. Annual rainfall ranges from 800 to 1200mm, while potential reference evapotranspiration (ETo) is between 1300 and 1500mm. The average annual rainfall can exceed 1500mm (closer to the coast), making it a sub-humid rainy climate. In terms of average annual temperature during the growing season, there are regions where temperatures exceed 25°C, although in general the average annual temperature varies between 20 and 25°C.

A considerable part of the interior has an altitude of between 200 and 500m, with undulating terrain that is sometimes interrupted by inselbeng rock formations. The district is crossed by several important rivers, which are not navigable due to their rugged course, but which contribute to agricultural activities and artisanal fishing (Mãe, 2005).

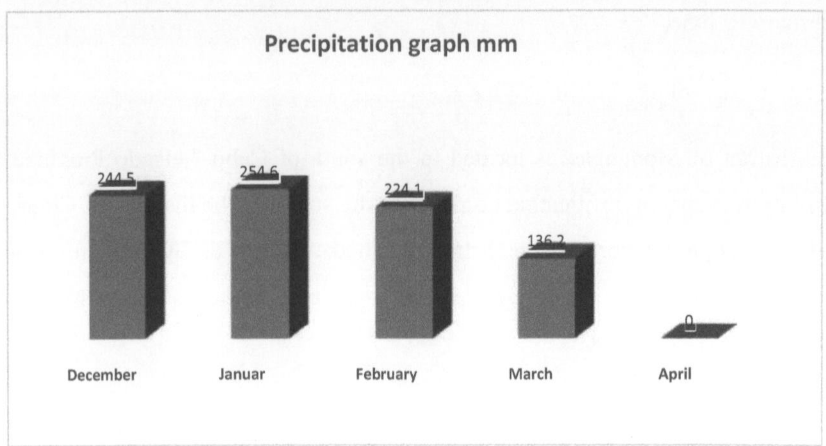

Precipitation graph for the Montepuez-SDAE-Montepuez district (2023)

3.4. Articles

For this study, we used sesame seeds of the Lindi variety, purchased from the farmer's house, with a minimum purity of 99%, germination of 80%,5, 4 botanical insecticides for the control of the sesame leaf beetle (*Alocypha bimaculada* Jacoby) based on castor leaves, margosa, eucalyptus and tobacco.

When preparing the botanical insecticides, extracts of fresh leaves crushed in a mortar (1,250g) were used for castor, margosa and eucalyptus. They were then placed in a plastic container (bucket) with five liters of water, where 250g corresponds to a liter of water for each botanical insecticide mentioned above. It was left to settle for one day (24 hours) and filtered through a clean cotton cloth to obtain a solution ready for use in the field.

Table 2. Treatment condition under study

Treatment code	Common name	Scientific name
T1	Control	*Control*
T2	Castor	*Ricinus communis L.*
T3	Eucalyptus	*Eucalyptus sp*
T4	Margosa	*Azadirachta indica L.*
T5	Tobacco	*Nicotine sp*

Source: Authors

3.5. Methods

3.5.1 Experimental design

For the present study, a scheme called Causalized Complete Block Design (DBCC) with a monofactorial scheme was used. This type of design is generally used when the place where the experiment is being carried out is not always homogeneous (Rossti *et all.*, 2017).It was conducted **on the** premises of the CIAM Centro de Investigação Agraria de Mapupulo, Cabo Delgado in the district of Montepuez, in an arrangement

- monofactorial consisting of four blocks (4) and five (5) treatments per block, for a total of 20 experimental units, the size of the plots corresponds to an area of 3m2 (1.50mx2m), total area was L=9m and C=15m which corresponds to 135m², Separation between blocks was 1m, in order to minimize error, the compass used was 0.60x0.10, I obtained a total of 3 lines per plot, 13 plants per line and a total of 78 plants per plot, two per cluster with a 15cm border on each line and 5 cm per plant.

3.5.2. Setting up the test

The trial was conducted in the 2023-2024 agricultural season under the conditions of the Mapupulo Agricultural Research Center, located in the

Montepuez district, and lasted four months. However, the following activities were carried out during the trial:

3.5.3. Land preparation

The land was prepared on January 30, and consisted of cleaning and manual plowing to a depth of 15 cm. The following day, the grass was removed and some of the logs that were in the field were removed.

2.5.4. Demarcation and sowing

The trial was demarcated and sown on the 1st of February, using ropes, a 100 m tape measure, 30 inch stakes and Lindi seeds. The compass used was 60 x 15 cm and (3) rows were obtained per plot and 13 plants in each row, making a total of 133 plants per plot.

2.5.5. Thinning

This activity was carried out after the plants had reached a height of 15 cm, in the third week of February after sowing. It consisted of reducing the plants in order to acquire two per plant, and was done manually, opting for plants that showed low vigour in terms of development.

2.5.6. Phytosanitary control

For the control, six sprays were made in February, March and April. The first was on February 10th, the second on the 21st, the third on March 3rd, the fourth on the 13th and the fifth on the 5th.

23 and the sixth on the 3rd of April respectively. Two 2-liter sprayers and one 20-liter dorsal sprayer were used, depending on the treatment, and then the material was washed and disinfected so as not to leave any residue or generate errors.

Table 3: Botanical insecticides used

Treatments	Insecticide Names Botanicals	Dose/ha	Dilution
T1	Control	-	-
T2	Castor	250g/ha	1 liter of water
T3	Eucalyptus	250g/ha	1 liter of water
T4	Margosa	250g/ha	1 liter of water
T5	Tobacco	3 dried leaves for 1 liter hot water	5 liters of water

Source: Authors

2.5.4. Sacha

Four weeding sessions were carried out in February, March and April. The first was on February 25, the second weeding was done on March 7, the third on March 20 and the last was done on April 9, with the aim of keeping the field clean and free of weeds that at some point compete for nutrients and can slow down the development of the sesame crop and at some point become vectors for pests.

2.5.5. Harvesting

The crop was harvested after it had reached its harvesting point, i.e. the physiological ripeness of the grain, which occurs when the pods begin to show a yellow color, then the plants were cut and tied into bundles and then subjected to a drying process in a free environment exposed to the sun's rays in order to continue the drying process of the pods for 25 days. After drying, the grains were threshed, sieved and weighed.

2.5.7. Data collection

The data collected during the trial included: herbivory index, pod weight, number of pods per plant, number of live and dead pests, yield per plot (kg/ha).

2.5.8. Herbivory index

For each treatment, 10 plants were harvested from a 3m² area and 20 leaves were selected from each individual, giving a total of 200 samples.

The percentage of herbivory was estimated on 200 leaves and the leaves collected were quantified with the naked eye. Dirzo and Dominguez (1995), where the leaf area consumed was classified according to the following categories:0-5%, 6-25%, 26-50%, 51-75%, 76-95% and 100% i.e. leaf area consumed ranges from 0 to 100% loss (Dirzo & Dominguez, 1995). Based on these class distributions and the quantification of damaged leaves, the following estimate of leaf damage was used:

Herbivory Index:

IH $\sum(n\text{ixi})/N$ =

Where: ni= number of leaves per category i= category per herbivore N= number of leaves for each sample area

To compare the herbivory indices between the different treatments, ANOVA analysis of variance was applied, followed by Tukey's test at a 5% probability of error level, using the Statistix10.0 package (Dirzo & Dominguez, 1995).

2.5.10. Weight of capsules

To achieve this, the capsules from each plant were selected according to the number of samples that were selected during the test, and the capsules were weighed using a portable precision scale purchased from the Mapupulo agricultural research center (CIAM).

2.5.11. Number of capsules

For this parameter, the plants were first selected at random, then the pods on each plant were counted and the total number of pods per plant was obtained.

2.5.12. Number of live and dead pests

In order to determine these variables, it was decided to observe and count the pests during the application of the botanical insecticides, i.e. before the application, the total number of live pests was counted, and after the pesticides were applied, the dead pests were counted in order to observe the performance of each botanical insecticide on the pest under study and the various pests that plague the sesame crop during production.

2.6. Income

During the sesame weighing process, the yield per plot was determined using a precision scale. The yield was determined using the formula.
Income formula

Yield (kg/ha) = □ grain weight x 10000 □2/useful area

2.7. Data analysis and processing

To analyze the data for the parameters in this study, the *ANOVA* statistical tool was used at 5% significance to determine the differences between the botanical insecticides. The Statistix 10.0 package was used to check for significant differences, and *Tukey*'s test was used to compare the means at 5% significance. The evaluation of the precision of the variation in the results obtained for the parameters under study was based on the coefficient of variation, in which if the CV is less than 10% it is considered low in terms of its classification and high precision, however, when the test shows 10-20% it is lower according to the classification and medium in terms of precision, and when the test shows 20-30% it is considered high in terms of the classification of CVs and low in terms of precision (Garcia, 1989).

Table 4: Classification of the coefficient of variation

CV values	Classification	Precision
0-10%	Low	High
10-20%	Minor	Media
20-30%	High	Low
30%	Very high	Very low

Source: Garcia (1989).

2.8. Constraints

The following constraints were observed during the study:

➢ Heavy rains in the first few days after sowing caused most of the seeds to be leached and germinate in unsuitable places, which led to the loss of some seeds, and following the same line, botanical insecticides were washed off after application, making the plants vulnerable to attack by pests and diseases caused by fungi.

➢ Shortages of some botanical insecticides, such as margosa, have led to people moving from one location to another in search of them.

CHAPTER IV: ANALYSIS AND DISCUSSION OF RESULTS

4.   ENTOMOLOGICAL RESULTS

4.1 Herbivory index

Graph 2: **Herbivory index of the pest** *Alocypha bimaculata* **Jacoby before and after the treatments**

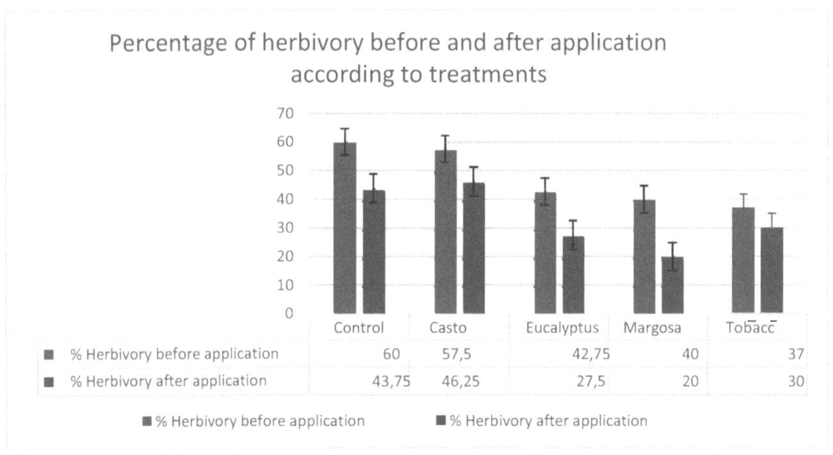

	Control	Casto	Eucalyptus	Margosa	Tobacc
% Herbivory before application	60	57,5	42,75	40	37
% Herbivory after application	43,75	46,25	27,5	20	30

Based on the data collected on the rate of herbivory in the field through the application of botanical pesticides, there was a significant difference in the rate of herbivory between the different treatments.

The graph in terms of the herbivory index shows that margosa had the highest average herbivory rates, where it lost 50%, followed by control with an average of 60%. This may be influenced by the climate in the region, where they invest more in growth and less in defences against herbivory, which results in greater attacks by herbivores (Coley & Barone, 1996).

In relation to climate, Turratti (2010) also carried out a similar study in a forest and found the same results. This increase in herbivory index rates in winter may be associated with low temperatures and a reduction in daylight hours, increasing the activity of herbivorous insects (Mari & Galassi, 2010). According to Dermacon et al. (2004), the rate of herbivory is lower in environments with greater availability of light, demonstrating that the rate of herbivory is directly linked to the incidence of sunlight.

However, based on the analysis carried out, it should be noted that there was a significant difference in the
concentration of the aqueous extracts on the percentage of herbivory by the leaf beetle (*Alocypha bimaculata* Jacoby) verified in the Tukey test carried out at a 5% probability of error for the castor, margosa, eucalyptus and tobacco extracts.

From a general perspective on botanical insecticides, the percentage of herbivory depending on the treatment shows a difference in the way the graph is read, with tobacco losing 7% of consumption, followed by castor oil with 11%, eucalyptus with 15%, the control with 16% and finally margosa with 20%. Translating this shows that in terms of the efficiency of botanical insecticides, tobacco was better than the other treatments.

Graph 3: Number of live and dead pests according to treatments

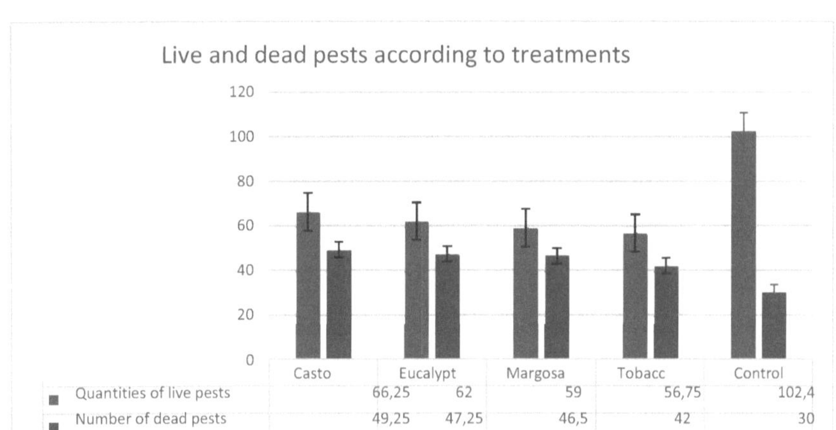

According to the graph of the number of live and dead pests, when the botanical insecticides was applied, it was statistically observed that castor oil had the highest average mortality rate, and the control had the lowest average mortality rate. A similar study was carried out by Jacomini et al. (2016), in which the beetle is sensitive to the insecticidal action of castor extract, achieving a mortality rate of over 30%.

According to Koppad & Shivanna (2010) nicotine is the main compound found in the extract, which can affect fertilization, insect behavior and cause toxic effects in insects. Nicotine has the power to act on the pest's central nervous system, binding to receptors (Sparks & Nauen, 2015).

For Margosa, an average of 46.5% mortality was obtained, since the active substance in margosa is *Azadirachtin*, which is effective in controlling other pest species. These results were obtained by Golo et al. (2013) when they observed that neem or margosa did not respond better in controlling *Alocypha bimaculata* Jacoby.

The Eucalyptus extract showed an average mortality rate of 49% for the pest under study, as with the other pests; this result is related to the extraction method used by Marconi et al. (2009). Based on the use of different aqueous botanical insecticides to control *Alocypha bimaculata* Jacoby. They found that castor oil had no effect on the pest under study.

However, based on the analyses carried out, it should be noted that there was a significant difference in the concentration of the aqueous extracts in the percentage of mortality of the leaf beetle (*Alocypha bimaculata* Jacoby) verified in the Tukey's test carried out at a 5% probability of error for the extracts of castor, margosa, eucalyptus and tobacco, which proved to be more effective in controlling the leaf beetle.

Similar results were found by Marques et al., (2013) who tested different concentrations of neither extracts having a significant effect on the mortality rate above 7%. However, tobacco extract, regardless of concentration, proved to be more effective in killing the pests, compared to the other extracts and concentrations evaluated. This is related to the active substance that tobacco contains (nicotine), an alkaloid derived from various plants, in particular *Nicotina tabacum sp*. It is a toxin that acts on the nervous system of pests and is very fast-acting (Menezes, 2005), according to the same author. Considered one of the most toxic botanical insecticides, it is used to control aphids, beetles, leafhoppers, cowbells, mealybugs and even crickets on fruit plants.

Agronomic results.

Table 5: Comparison of means according to the number of capsules per plant

Treatments	Media
Tobacco *(Nicotina tabacum sp)*	39.000 A
Margosa *(Azariactina indica)*	33.000 A
Eucalyptus *(Eucaliptus sp)*	32.500 A
Castor *(Ricinuscomunis L)*	22,750 AB
Control	13.000 B
CV(%)	7.34
General media	55.850
Probability	0.0135

Means with equal letters do not differ significantly from each other according to Tukey's test at 5% probability.

With regard to the effect of applying botanical insecticides on the parameter analyzed, Tukey's test at the 5% level showed that there was a significant difference.

With regard to the number of capsules per plant, the lowest average was 13,000, obtained in the control (without application), while the highest average was 39,000 obtained with the application of tobacco extracts, representing a greater increase in growth of 75% Silva et al. (2016).

Ávila and Graterol (2005) also found a significant effect of using organic products on the number of pods per plant. The number of pods is directly linked to the productivity of the sesame plant, as is the emission of productive branches (Severino, et al., 2002).

Mesquita (2010) in his study also found an average value of 143 capsules, which is higher than the value found in this study. Vieira et al. (1994) reported that sesame production under rainfed conditions, following all the requirements of the crop, can produce up to 70 capsules per plant. Beltrão *et al.* (1994) also reported that

Using different ways of planting three varieties of sesame, they state that they obtained average values ranging from 102 to 135 capsules per plant.

Weight of capsules

For this parameter, the capsules were weighed with a precision scale for all treatments, and the results can be found in the table below. The lowest average was 35.250 and the highest average was 66.250.

Table 6: Comparison of means according to weight of capsules per plant

Treatments	Media
Tobacco *(Nicotina tabacum sp)*	56 A
Margosa *(Azariactina indica)*	59 A
Eucalyptus *(Eucaliptus sp)*	62 A
Castor *(Ricinus communis L)*	66, 250 AB
Control	35,250 B
CV (%)	7,94
Overall average	42,600
Probability	0,0772

Averages with the same letters do not show a significant Tukey's test at a 5% probability level.

The results obtained in the analysis of variance at the 5% probability level of error, Castor extracts showed a higher average weight of capsules per plant when compared to the other varieties as well as the control.

Magalhães et al., (2010) observed statistically significant effects between treatments on the weight of pods produced in sesame grown organically with cattle manure under semi-arid conditions in the region located in the Parabian hinterland, with a tendency for the average to increase as the doses of cattle manure increased, as its highest value was verified with a dose of 40 tons/ha. The same value disagrees with a study by Alves (2014) in which the author observed a 69% increase in the weight of pods per plant when he applied 12 liters of manure per linear meter. However, a decline was observed in the tobacco and post-control treatments, none of which matched the values observed in the different treatments, with the exception of castor with an average value of around 66.250. The other treatments were not efficient in terms of pod weight per plant in this experiment due to the following external factors: water absorption capacity, filling, heterogeneity of the environment and sowing date **Table 7: Yield in kg/ha**

Treatments	Media
Tobacco *(Nicotina tabacum sp)*	1501,0 A
Margosa *(Azariactina indica)*	992.5 AB
Eucalyptus *(Eucaliptus sp)*	743.3 BC
Castor *(Ricinus communis L)*	660.0 BC
Control	267, 5 C
CV (%)	35,13
Overall average	832, 85
Probability	0,0010

Averages with equal letters do not differ significantly according to Tukey's test at 5% probability level

Based on the results obtained in the *ANOVA*, significant differences were observed between the treatments, where according to the *Tukey* test, the treatment applied with *Nicotina tabacum sp* extracts was the one that showed the highest average d1501.0 kg/ha compared to the other treatments and the lowest average was observed with the *Castor* extracts with 660.0 kg/ha.

The results found above in Table 5 are not in line with the results obtained by Lima (2011), who in an area of 15m², using sesame with a spacing of 10 cm between plants and 60 cm between rows, with only 70 capsules per plant, estimated a population of 400,000 plants per hectare and obtained a yield of 2,292 kg/ha. It can be clarified that this figure is very high since the experiment was carried out on a very small area, but he says that if the experiment had been carried out on a full scale, the control over soil and plant management would not have been as effective, thereby reducing productivity.

Mesquita (2010), through studies carried out in a greenhouse with the sesame crop, estimated a yield of 1000 kg/ha, a lower value when compared to the present study. Perin *et al.* (2010), based on field experiments, obtained a yield of 843.43 kg/ha, i.e. there was a significant difference, and the data obtained in this study showed a higher yield compared to the studies carried out by the aforementioned authors.

However, the results obtained in this study are not in line with those reported by the IIAM (2020) where the average yield of sesame varies between 800 and 1000 kg/ha for the northern and central regions of the country (Mozambique).

The coefficient of variation obtained for sesame grain yield (35.13) is considered to be of low precision (Garcia, 1989).

Conclusion and recommendations

Based on the effectiveness of botanical insecticides in terms of controlling the pest under study, the leaf beetle (*Alocypha bimaculata* Jacoby), tobacco *(Nicotina tabacum sp)* was the best performer with a yield of 1501 kg/ha, followed by *Azadirachta indica* with 992.5 kg/ha.

There were statistically significant differences in the parameters studied (number of capsules and weight of capsules). With regard to the parameters of herbivory index, number of dead and live pests, there was a significant difference in the Tobacco treatment, while the other treatments showed no significant differences in these parameters. In terms of production, margosa showed better performance in controlling the other pests, while tobacco was efficient in controlling the pest under study. Therefore, the yield of *Nicotina tabacum sp* extracts had a higher average of 1501.0 kg/ha. Therefore, the null hypothesis (H0) is rejected and the alternative hypothesis (H1) is accepted.

Recommendations

Based on the results obtained in this study, the following is recommended:

➢ To the IIAM - Instituto de Investigação Agraria de Moçambique to carry out similar studies in this or other places with botanical pesticides based on extracts of margosa and tobacco in order to convince the population to adopt new production techniques without having to resort to the use of chemical products since they cause damage to the environment and to human health.

For sesame growers it is recommended:

In terms of production, margosa proved to be efficient in controlling primary and secondary pests through the treatments, but in terms of controlling the pest under study, the leaf beetle (*Alocypha bimaculata* Jacoby), I recommend the use of extracts based on tobacco, so the producer is called upon to make a small effort to better understand the treatments and adopt new production techniques that are sustainable and easy to acquire.

BIBLIOGRAPHICAL REFERENCES

Antoniassi, R., A, Gonsalves, (2013) *influence of cultivation conditions on the composition of sesame seed oil.* V 60 no.3.

Andrade, L. N., & Nunes, M. U. (2001). *Alternative products for disease and pest control in organic agriculture.* Beira-Mar: Embrapa Tebuleiros Costeiros.

Arriel, N. H., Firmino, P. d., Beltrão, N. E., Soares, J. J., Araújo, A. E., Silva, A. C., & Ferreira, G. B. (2007). *The sesame crop.* Brasilia, DF: Embrapa.

Beltrão, N.E.; F. freire, E.C.; Lima, (1955) E.F, *Gergelim cultura no tropico semi- arido.* Campina Grande: Embrapa-CNPA, 52p.

Beroza, M.; kinman, M.L, (1955) *semain, sasamondin and sesamol content of the oils of sesame as affected by strain, location graown, agring and frost danage.* Il v,32, p.348-350.

Biondie, M., & DAlessandro, P. (2011). Jacobyana Maulik, *a genus of Oriental flea beetle new to the Afrotropical region with description of three new species from Central and Southern Africa (Coleoptera, Chrysomelidae, Alticinae).* Zoochaves, 47-59.

Carneiro, J. S. da S., Silva, P. S. S., Freitas, G. A. de, Santos, A. C. dos., Silva, R. R. da.

(2014) *Response of sesame to fertilization with bovine manure and doses of phosphorus in southern Tocantins.* Revista scientia Agraria, vol 17 n 2, p 41-48.

Clusa, (2008). *Extension worker's manual for sesame production.* Clusa Nampula. Coley, P,

D. & J, A., Barone (1996) *herbivory and plant defenses in tropical forests.*

*Annual Rev.*Ecol. Syst. 305-335.

Cloyd, R. (2004) *Natural indeed: are natural insecticide safe and better then conventional Insecticide? Illinois pesticide review,* 17: 1-3.

Dirzo, R., Domingues. (1995). *Plant-herbivore interaction in mesomerican tropical dry*

Forest in: bullock, S. H., Moone, H, A., Medina, E. A, (eds). Seasonally dry tropical Forests. Cambridge: Cambridge University Press, P.304 - 325.

Embrapa Brazilian Agricultural Research Corporation, (2014) *Sesame cultivation,* 2 ed,londrina.

Embrapa, Embrapa soils (2009). *Manual de análise Química de* solos *e fertilizantes*, Brasília: Embrapa solos.

Ferreira, P. G., Huther, C. M., Carvalho, A. S., Forizi, L., d., Silva, F. d., & Ferreira, V. F. (2020*)*.Nicotine and the origin of neonicotinoids: problems and solutions. *Virtual journal of chemistry, p. 403.*

Fonseca, K.S. *evaluation of the physiological ripeness of sesame (sesamum indicum L)*

To obtain the ideal harvest point. Campina Grande Federal University of Paraíba, 1994. 35p. Supervised internship dissertation.

Frandoloso, A., (2006). *Growth and productivity of the sesame crop (sesamum Indicum L.) under different fertilization levels.* Botucatu, v 18, n.

2, p. 364-375.

Garcia, C.H., (1989). *Tables for classifying coefficients of variation. Piracicaba*: IPEF, (Technical circular, 171) 12p.

Guimarães, Rita de Cassia (2012*). Sesame* seeds *(sesamum indicum L.)* oil effects on serum lipids and glucose large Ms fields.

Grilo, J. J., A. S., Azevedo, P. V. de (2013) Growth, development and productivity of BRS Seda sesame in Agrovila de Canudos, in Ceara Mirim (RS). Revista Holos, v.2, P.19-33. n2, p363-369.

Lima, F. V., Pereira, J. R., Araújo, W. P., Almeida, S. A. B., Leite, A.G., (2011). *Defining spacing for irrigated sesame.* Revista Educação Agrícola Superior, ABEAS, v 26, n.1, p 10-16.

Linzmeier, G. R., and Ribeiro-costa. V. V., (2012). *Review of the genus blosyrus schoenherr (coleoptera, chysomelidae) from Tanzania.* Oriental insect, p. 29.

Koppad, G. R., N. Shivanna (2010). *Effect of nicotine on larval behavior and fitness in drosophila melanogaster.* Journal of biopesticides, v 3, p. 222-226.

Mãe, (2005). *Perfil do distrito de Montepuez província de Cabo Delgado.* Ed.

Mozambique: Ministry of State Administration.

Magalhães, I. D., Costa, F. E., Alves, G. M. R., Almeida, A. E. S., Silva, S. D.; Soares, C. S. (2010).

Organic sesame production under semi-arid conditions. In: IV Brazilian Congress of

Mamona & I Simpósio Internacional de oleaginosas energéticas, João possa- PB, 2010.
Proceedings

Campina Grande: Embrapa Cotton, p. 749-754.

Martinez, S,S. (2002),o neem: *Azadirachta indica, usos múltiplos, produção londrina:*

IAPAR. 142p.

Marconi, A. M., Alves, L. F. A., Bonini, A. K., Mertz, N. R., dos Santos, J. C. (2009).
Insecticidal activity of plant extracts and neem oil on adults of beetles (coleoptera). Arquivo
instituto biológico, *v. 76, n. 3, p.409-416.*

Marques, C. R, G., Mikami, A, Y., Pissinati, A., Piva, LL. B., Santos, O, J. A. P., Ventura,

M. U, (2013). *Mortality of beetles by neem and citronella oil.* Semina: ciências agrarias v34,

n 6, p2565-2574.

Menesse, P. R., & Almeida, T. d. (2012). *Reseachgate.* Retrieved from
https://drive.google.com/

/file/d/1wphyfrboxowhnr6lfov-xQMZ-_JIZHa/view.

Mesquita, J. B. R. (2010). *Management of the sesame crop subjected to different irrigation*

rates, doses of nitrogen and potassium by the conventional method and by ferti-irrigation.

Dissertation (master's degree in agronomy) - Federal University of Ceara, Fortaleza.

Milani, M., Junior, S. R., & Sousa, R. d. (2009). *Sub-species of castor bean.* Campina grande,

PB: National Cotton Research Center.

Moreira, M. D., Picanço, M. C., Silva, E. M., Moreno, S. C., & Martins, J. C. (2017).
Use of botanical insecticides in pest control. Brazil.

Morreira, M. D., Picanço, M. C., Silva, E. M., Moreno, S. C., & Martins, J. C. (January

2007). Use of botanical insecticides in pest control.

Monteiro, A. F, Pereira, G. L., Azevedo, M. R. Q. A., Fernandes, J. D., Azevedo, C. A, V.

IIAM, (2011). Annual activity report agricultural campaign

2009/2010. IIAM, (2020). Quarterly report on the 2018/2019 agricultural season.

IITA, (2012). *Extension worker's handbook.*

IAPAR, (2006)   *InstituteagronómicodoPanará.*   http:/www.iapar.br/zip

pdf/nim2.pdf. (February 14).

Iria, M., José, R., Guedes, J., Katell, & Sonia (2020). *Natural pesticides: Alternative management for pests and diseases.* Brazil: CIP-Brasil.
Jacobson, M. (1989) botanical pesticides: past, present and future. In: Arnason, J. T., Philogene, B. J. R., Morand, P. Insecticide of plant origin. *Washington, DC, American Chemicals Society.* V.387, p.69-77.
Jacomini, D. Temporini, L. G., Alves, L. F.A., Silva, E. A. A., Marinho, T. C. J., (2016).

Tobacco extract in the control of the poultry beetle. Pesquisa agropecuária brasileira, v 51, n.5.

Jolivet, S. S., and Verma. H., (2002), Vareital suceptibilly of (*sesame sesamum indicum L*)

Alocypha bimaculata Jacoby. JNKVV Res J., 8 P. 3-4.

Júnior, J. A. S., Azevedo, (2013), *growth and development productivity of sesame.* Revista hools, v,2, p.12-33.

Júnior, J. E. P., Santarosa, E., & Goulart, I. C. G. R. (2014). *Growing eucalyptus on rural properties:* diversifying production and income. Embrapa. Brasília, Df.

Lago, A.A; Camargo, O., B. A.; Savy e filhos, A; Maeda, J, A (2001). *Maturation and seed production of sesame cultivar IAC-china pesquisa agropecuária Brasileira,* Brasília 36, (2014). *Hydroponic cultivation of sesame cultivars in organo-mineral nutrient solutions optimized with the* Solver *tool.* V 18, p. 417-424.

Neves, E. J., Carpanezzi, A. A., Vianna, P. A., Ribeiro, P. E., Prates, H. T., Malimpence, R. A.,Santos, A. J. (2008). *The neem crop.* Brasilia: Embrapa Informações Tecnológicas.

Neto, M. E., Pereira, W. E., Souto, J. S., Arriel, N. H. C. (2016) *Growth and productivity in Fluvic neosol as a function of organic and mineral fertilization.*

Rer. Ceres vol 63 n 4 viçosa Jul/ Aug.

Oliveira, Jason Silva, (2000). *Situation of sesame in Barreiras.* Barreiras de Bahia.
Olmi, (2006), v3: p183 Integrated pest management.

Palote, D. S S., (2017). *Review on beetles (coleoptera): An Agriculture major crop pets the words. International jounal of life-sciences scientific.* Research 36, 1424-1432.

Paulo Afonso Viana, P. E. (2010). Effect of aqueous extracts of green neem leaves (*Azadirachita indica*) and time of application on damage and larval development of Spodoptera frungiperda (J.E. Smith, 1797). *Revista Brasilleira de Milho e Sorgo, v.9,n.1,p,* 27-37.

Perin, A. D., J. Silva, J. W., (2010) Sesame performance as a function of NPK fertilization

and the level of soil fertility. *Actascientiarum Agronomy*, v. 32, n. 1, p. 93-98.

Picanço, M. C. (2010). *Integrated pest management.* Brazil: Viçosa, MG.

Queiroga, V. P., Arriel, C., Silva, (2010) *technology for the sesame agribusiness.* *Embrapa* cotton 1 ed.

Queiroga, V, P.; Silva, O. R.R.F, (2008), *technologies used in mechanized sesame cultivation.* Embrapa, p142.

Rech, T. (2018). *Diversity of Alricini (Newman, 1834) (Coleoptera, Chrysomelidae Galerucinae) in forest fragments in southeastern Panara.* Brazil: Dourados- Ms.

Luciano S. S.Z. N. (2008). *Reachgate.* Retrieved from ainfo.cnptia.embrapa.br

Shiratsuchi, L. S. (2008). *Reaseachgate.* Retrieved from ainfo.cnptia.embrapa.

Silva, N. F., Nascimento, L. F., Santos, R. F., Junior, L. A., Cunha, E., & Rocha, E. O. (2019). Characteristics and cultural treatments of *sesame (sesamum indicum)*. *Revista Brasileira de Energias Renováveis v8, n.4*, 665-675.

Sparks, T. C., Nauen, R. Irac; (2015). *Mode of action classification and insecticide resistance management.* Pesticide biochemistry and physiology, v 121, p 122- 128.

Turrati, M. P., (2010). *Distribution patterns of herbivorous insects in a fragment of dense montane ombrophilous forest in southern Santa Catarina.* TCC (Undergraduate degree in biological sciences) - Universidade do Extremo Sul Catarinense, Criciúma, 43 f.

Vilela, J. A. (2008). *Effect of Using Neem Oil (Azadirachta indica) Dermally and Moxidectin Subcutaneously to Prevent Artificial Infestations by Dermatobia hominis (LinnaeusJr., 1781) (Diptera: Cuterebridae) in Cattle.* Rio de Janeiro.

I want morebooks!

Buy your books fast and straightforward online - at one of world's fastest growing online book stores! Environmentally sound due to Print-on-Demand technologies.

Buy your books online at
www.morebooks.shop

Kaufen Sie Ihre Bücher schnell und unkompliziert online – auf einer der am schnellsten wachsenden Buchhandelsplattformen weltweit! Dank Print-On-Demand umwelt- und ressourcenschonend produzi ert.

Bücher schneller online kaufen
www.morebooks.shop

Printed by Books on Demand GmbH, Norderstedt / Germany